Superseed!

Superseed!

Tom Stanks

PENTLAND PRESS, INC.
ENGLAND • USA • SCOTLAND

PUBLISHED BY PENTLAND PRESS, INC.
5122 Bur Oak Circle, Raleigh, North Carolina
United States of America
919-782-0281

ISBN 1-57197-087-8
Library of Congress Catalog Card Number 97-069239

Printed in the United States of America

Foreword:

This is a story for travelers of all ages. Change the names and situations, and I believe everyone is a superseed going toward fuller life, integration and oneness. A child's unspoiled imagination grasps this. With a few nudges here and there on the way, maybe adults can too. This is a story to read and live, and hopefully, to read some more.

Acknowledgment:

Thank you, Melvin Powers, for planting a seed. I am indebted to you, Nature, for revealing your secrets to me. I am forever grateful to you, my spirits and guides, for teaching me a new language to express truth and beauty together. Tuesday Breakfast Group, thank you for the abiding support and friendship that I rarely find in a gathering of men. Jim Irwin, you have a touch for the right word. Your fine editing removed inaccuracies and repetitions and kept the narrative moving forward. Jack Vazquez, you saw some faults that could have hurt the account's authenticity. Marty Peters, you have done so much so well. Thank you for the hand you have extended in the course of our friendship, but especially for helping me shape the text into a delightful telling. Jim Griffin, your agreement on content and your conviction that the book will be read and reread gave me the final go ahead. Michael DeGuire, your suggestions for the artwork and help in promoting the book were invaluable. To you, Helen, my wife, companion, aide, I am beholden for enchanting the story. This book would not have been possible without you.

awakening . . .

Chapter One:

stuck

Once upon a time, in a land just like this, there lived a little fellow who was searching. Searching, searching, searching . . . not for gold and not for riches. Nor did he want to slay dragons, save fair princesses or be a great hero. His quest was inside himself. He wanted to know who he was. He had tried many times to find out, all to no avail. It seemed he was stuck.

In fact, he was stuck in some high grass under a tree. Near the top of the grass blade, he could see the black earth below him and the other grasses blowing in green waves around him. When he looked up, he found an enormous tree bent over him almost blocking out the sky. He asked himself, "How did I ever get stuck here?" He tried many ways to free himself but nothing worked. He felt miserable.

He knew his name was Vol but wondered, "Is this all I'm going to know about myself? Why, I don't even know how I got this name." Everything seemed wrong. He wanted to get hold of himself but felt helpless. Finally, tired and frustrated, he began to cry. With his crying, a tear fell and it carried him to the ground. He shook to get the water off and realized he had been freed. Suddenly he had a fresh outlook. He decided to make friends with the teardrop.

"Who are you?" asked Vol.

"My name is Plu," answered his new companion.

"What do you do?" Vol asked.

"I water," replied Plu.

"Hmmm! You do laundry?" questioned Vol.

"I wash with water," answered Plu. "What is your name?"

"I'm Vol. That's one thing I do know. But there are so many things I don't know. Like what am I doing here, or how did I get hung up on grass? I think I'd still be there if you hadn't dropped by."

"Well, I didn't just drop by, Vol. You called me with your cries. As to your question on being hung up, it was the wind that blew you to the high grass. The other question, Vol, is more difficult. In fact, I don't think you can find out all at once what you're doing here. But if you want, I can help you get started."

"That's great. What will you do?"

"No, Vol. What will *you* do? I'm just going to help. To progress, you will have to trust me."

Vol preferred to do things on his own and didn't like having to trust others. On the other hand, he realized he had been stuck until Plu came along and woke him. Furthermore, he still didn't know his purpose. Quietly he said, "I think I can trust you. Tell me what to do."

"OK, I'll be your companion and guide on your journey. Sometimes, however, you won't know I'm around, and you'll feel alone. Do you think you'll be all right when I'm not there?"

"Oh sure," answered Vol, much too quickly, his idea of independence reasserting itself.

"All right, let's get started," said Plu. "Do you like it around here?"

"What do you mean, around here?" questioned Vol.

"Do you feel at home on this part of the earth?"

"Well, I like being back on the ground. Yeah, this feels cool again."

"Hmmm! Soft and fertile," muttered Plu, half to herself as she felt the ground.

"What did you say?" asked Vol, drawing closer to Plu.

"I said there's nice topsoil here. We can really get into it now if you like."

"Yeah, I'm ready. I've been hung up long enough. What do I do?"

Plu told him, "I'm going to work you down below the surface into the darkness of the earth. You'll feel my presence for a while, then some new feelings will come over you."

"New feelings? What kind of feelings?" asked Vol, getting nervous again.

"Let's just say they are necessary experiences. You still want to know who you are, don't you?"

Again Vol felt stuck—stuck unless he'd commit himself. But he still wanted an easy way out. "Can't you just tell me what will happen and we can get on with it?"

"No, Vol. Then it wouldn't be yours. Your story doesn't come from someone else telling it. You have to live it and in living, create it yourself."

"Well, how long will I be down there?" asked Vol. "Will it be a long time?"

"Mostly it depends on you and to a lesser extent on the conditions down there." Plu wanted to impress upon Vol that he had to experience it for himself. "All the time you'll be discovering yourself. That's what you want, isn't it?"

"Discovering sounds good to me," answered Vol. "I'm ready to start."

"Well, Vol, if you're ready, just relax and let go. For now you don't have to do anything but skim along with me. You know, go with the flow. When you come to a place that feels comfortable, you'll act more on your own to forward your life."

In what seemed like magic, Plu reached out to join up with some of her traveling companions. Together they took Vol on the voyage of his life.

Chapter Two:

underground

"Gee, it's dark down here," said Vol. "Would someone turn on a light? This place is like a tomb."

"You don't need light—outside light, that is," answered Plu. "It would just blind you to the work to be done."

"It's also so quiet," continued Vol, missing Plu's distinction. "What work are you going to do, Plu?"

"No, Vol. What work will *you* do? You see, this tomb, as you call it, is more like a womb. It's meant to give life. The darkness and silence are here for your connecting."

"Connecting with what?"

"With everything. We're all connected, Vol, with each other, with the wind, with the sun, with the sky, but we don't realize it. That's what you're here to become aware of, and once you know with your heart that we're all connected, it helps you to see that we're all one. You need to find the deep center of silence within yourself so you can connect. The connection brings you face to face with others without anything in between."

"What happens when I connect?"

"Your existence spreads like water,"

"My existence spreads like water?" repeated Vol.

"Yes, that is, your true self gets bigger," explained Plu. "In that deep silent center is your potential. It makes you able to listen. Remember, you said you wanted to find who you really are."

"Yes," replied Vol, "but if I'm to find out who I am, what do others have to do with it?"

"They help make you what you are. It's like being blinded by the light I mentioned earlier. When your own thoughts are too loud, you can't hear what others say. You like to be heard, don't you?" Vol nodded. "Well, others like it, too. You can't find yourself when you shut out others with your own loud thoughts. By being quiet, you open up to them. That will give you some sense of belonging, but just how you belong will be up to you."

"Boy, this is getting complicated," remarked Vol.

"It may seem complicated, but it gets simple," answered Plu. "Silence becomes complicated when words and thoughts get in the way of communication. It takes silence to listen.

"To put it another way, you have what you need within you, as everyone has. You just need to let go to find it. And just as you are to be fulfilled, so are they. But you can't be an obstacle to what others need. All of us are to be fulfilled. Some of us are just slower to pick up on it."

"So, what do I have to do—say my prayers and go to sleep?"

"No shut-eye down here," Plu said sharply. "It's more like a Quaker quiet than sleep. That silence I told you about is to keep alert. You're going to have to be aware of what is going on around you as well as within you so you know how to respond and find out who you are.

"Discovery comes from positioning yourself, Vol, that is, between what others say and what you say, between acting and watching."

"You make my brain sweat. You're telling me so much I haven't heard before. I suppose you expect me to take it on faith, since I just don't understand a lot of what you're saying."

"You can if you like. Eventually, your experience will replace any faith you may have put in me. Vol, you're on your own now. There's a good chance we'll meet later, but that's up to you. Take care."

* * * * *

Left alone, Vol tried to think of those with whom he might connect. "Who in this tomb will be my connection?" The more he thought about it, the more comfortable he became. And in his ease he sang, "*Whom in the womb of the tomb? Whom in the womb of the tomb?*"

He thought of how neat it had been to be with Plu. Then he had had someone to talk to, but now he had no one. "I wonder why I feel so lonely. But that's why Plu said I'm here—to connect so that I won't be lonely. But when I try to connect I still feel lonely. Oh boy! *Whom in the womb of the tomb? Whom in the womb of the tomb?*

"Is there something down here that will take my loneliness away? Will it lead me elsewhere to something that will fulfill me? Oh, I'm not getting anywhere. *Whom in the womb of the tomb? Whom in the womb of the tomb?*"

With nothing happening, Vol spent more time reflecting. "Plu said I don't just discover myself, I create myself by living. Some day I'll have to decide what I want to be. I guess that would be creating myself. *Whom in the womb of the tomb? Whom in the womb of the tomb?* Maybe my connection will help me make up my mind."

Time hung heavy and dark. Vol meditated, but he focused his attention on creating a dance to the tune, "*Whom in the womb of the tomb? Whom in the womb of the tomb?*" The tune made him feel good, and he strutted around, kicking out and enjoying himself.

Vol eventually tired and started feeling cold in the darkness. He remembered the light and wondered whether he would be warmer if he went up toward the light. Vol tried to lift himself but couldn't. He tried again and again but didn't budge. He got much colder and more tired. Vol didn't sing or dance. Instead, he got angry at Whom in the womb of the tomb.

"What do I do now?" he asked in desperation. He remembered that he hadn't taken himself to the high grass; the wind had. And how had he gotten down here? His buddy Plu had brought him. "If I'm so independent, how come I can't help myself?"

Vol pondered his predicament. "Maybe that's why I'm to connect. I'm in need and didn't know it. Well, I certainly know it now."

The need kept pestering Vol. "Air and water are simple things that everybody knows. Air blew me to the high grass and water brought me here. *Whom in the womb of the tomb? Whom in the womb of the tomb?*"

Then a thought came that hit Vol like underground lightning. "My goodness! Are water and air and light the things I'm supposed to connect with? I've taken them for granted all this time and I need them. In fact, maybe I can't move without them."

"Not only can't you move, you can't live without us," someone said.

"What? Who are you?" Vol asked, frightened by the thought that his life may depend upon others and overwhelmed because the voice seemed to come from all around him. He asked again, "Who are you?"

"Some call me Mother. I am the Earth that gave you birth," answered the voice.

"I am the Earth that gave you birth" reminded Vol of "*Whom in the womb of the tomb?*" "Where did you come from?" asked Vol.

"I was here all the time. You notice us when you become aware of how important we are."

"I'm sorry. I just never dreamed that I needed all of you so much, that we kind of go together."

"Not just go together, Vol. You must allow us to become part of you."

Now Vol really got frightened. He thought he might be buried forever and lose himself. "I like myself the way I am, thank you. Can't you just lend me a helping hand?"

"Yes, but sooner or later you'll have to open up to bigger things, like Plu talked about."

The mention of Plu made Vol think, "*Whom in the womb of the tomb? I am the Earth that gave you birth.*" "How did you know about Plu? You weren't there."

"We're connected. And being connected is what you came here to learn, remember?"

"Oh, yes. But how do you become part of me?" *Whom in the womb of the tomb? I am the Earth that gave you birth.*

"You just say yes to me and I become you. You won't die. Nothing really dies; it just gets changed over into its next phase. That's what I will do in you, and that's what you will continue to do."

"I will get changed over?" *Whom in the womb of the tomb? I am the Earth that gave you birth.* "How will I change?"

"Well, you'll go through phases where you move from one stage to the next."

"I will go through phases?"

"Yes, there are many steps in life. I guess no one ever told you. You have a sunny future. Would you like to get on with it?"

Whom in the womb of the tomb? I am the Earth that gave you birth. "I'd like to go toward the light so I can get warm."

"Great! That's a step in the right direction. Soooo . . . hold on. We're on our way."

Vol could feel himself being moved, as though in an earthquake. But he also was aware of his willingness to allow it to happen. *Whom in the womb of the tomb? I am the Earth that gave you birth.* He found himself saying yes to the air and water and now to the earth. It was a new experience for him, an accepting and letting go at the same time.

He felt himself a part of something bigger than himself. *Whom in the womb of the tomb? I am the Earth that gave you birth.* "So this is the connecting Plu told me about. Well, well, well. It isn't so bad after all. I guess Earth is a solid fellow, er, woman." Now Vol was starting to feel better, but he was still anxious over the phases he might have to face. He was afraid he would lose something of himself. *Whom in the womb of the tomb? I am the Earth that gave you birth.*

The warming gradually seeped through Vol and felt good to him. This new place also was brighter, and he liked that. *Whom in the womb of the tomb? I am the Earth that gave you birth.* He remembered that Earth told him she was always there, but he would become aware of her only when he needed her. He said to himself, "I guess the warmth and light are always here. I just need different things at different times. I certainly needed Plu. What about air? I need that all the time. Good grief! Have I been taking *everything* for granted?"

These thoughts caused Vol to feel great sorrow. He grieved after remembering how much others helped him—air, water, earth and light—and he hadn't realized it. He wondered whether the new phases would be as this one had been—one discovery and rediscovery after another. His sorrow increased and he cried for a long time.

After his energy was spent, he felt some relief. With time passing, he started to think about how he would conduct himself in the coming phases. "In the future," he vowed, "I will be more sensitive to my surroundings."

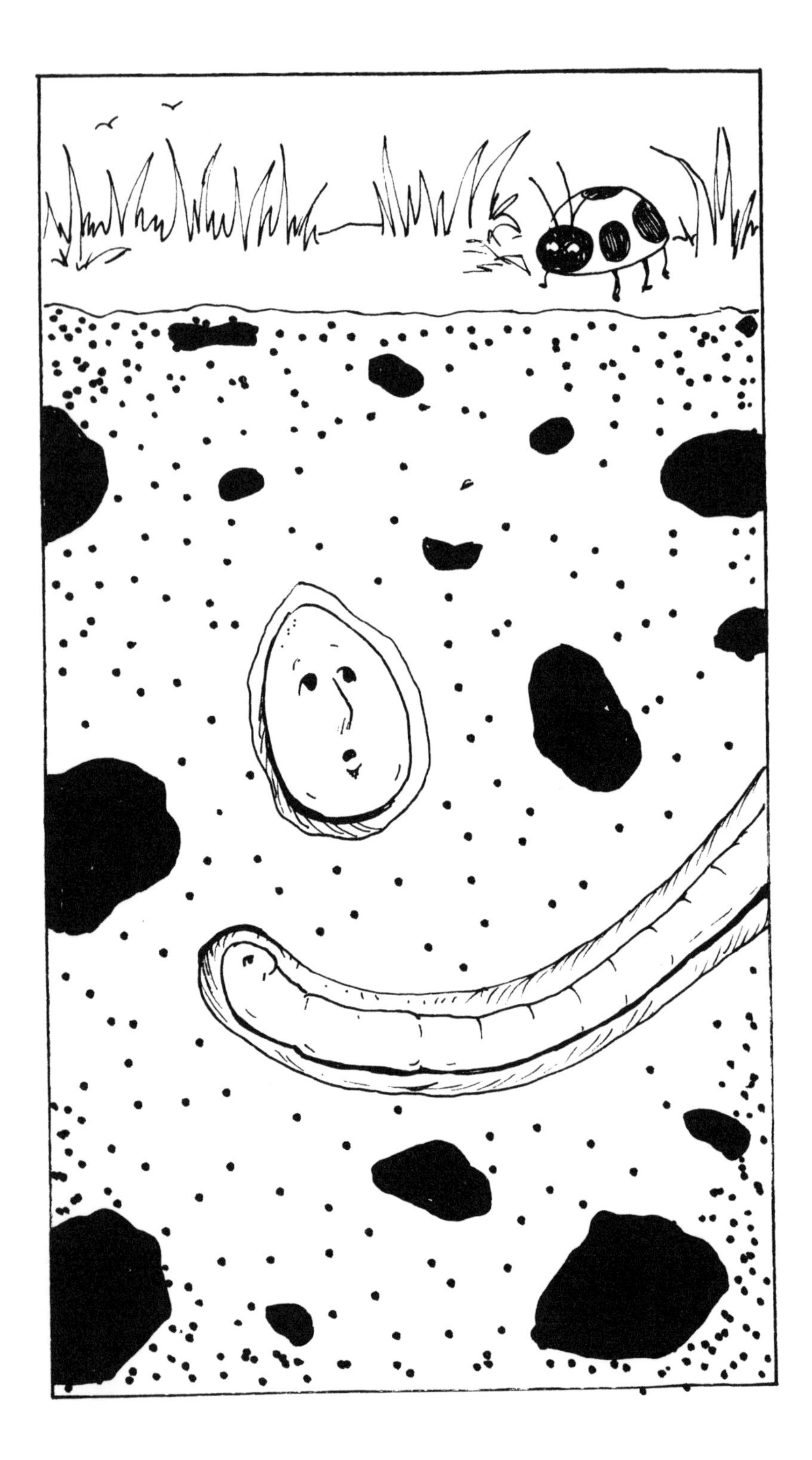

Chapter Three:

afterripening

A long time passed. Nothing happened. Vol wondered whether he had finished with the connecting that Plu told him about. "But I still feel the same. If this is a new phase, I sure can't tell," he thought.

Vol began to feel depressed. He wondered whether he was meant to stay like this while life passed him by. It was warmer and lighter, but excepting that, everything seemed the same. He began to have many doubts and sensed his inadequacy.

"Why do I feel so inadequate? I don't believe I'm meant to be helpless. Maybe I feel this way because I depend on others too much. Now I know I need them, but maybe I have to act on my own to get what I want."

He thought about Plu. He hoped they would meet sooner rather than later, but now it seemed to be getting very late. He remembered that she said they might meet again, but that it would be up to him. He longed for her company. "I've got to try something."

After all his self-talk, in desperation he finally squeaked out his longing, "Plu, if you can hear me, I wish you would come to me. I'm tired and I feel helpless. I don't know what to do. I linked up with you once, Plu, and I need you again. Are you there, Plu? Plu, can you hear me?"

"Yes, I'm here, Vol. You called me. You asked for me. So now you've connected with me," said Plu. "It just takes your intention and action to have me come."

"Plu!" exclaimed a relieved Vol. "I've thought about you so much. It's good to see you."

Vol was overjoyed by Plu's appearance and honestly interested in *her* rather than what he had been enduring. "Drop in on anyone recently?" he asked, trying to be funny.

"Just enough to keep the cycle going."

"I was expecting to enter the next phase of my cycle by my connecting, but nothing happened," Vol said, his frustration resurfacing.

"I thought you'd have moved on by now," said Plu. "Honestly, I don't know what's taking you so long. Can I ask you some questions? This is going to be personal. I hope you can keep in mind that I'm trying to help you. Are you ready?"

"Yes," replied Vol.

"Do you have any idea yet who you are or what your purpose is?"

"That's what I set out to do, Plu, as you know. And you've helped me. I felt so frustrated that I had to call on you. I know you put me on the right track. I'm trying to get closer to the answers by the connections I've made in silence. I feel a little less separate. But no, I still don't have the answers."

"We might have to vary the formula a bit and keep you down here longer. What do you think about that?" asked Plu.

"Did I do something wrong?" asked Vol.

"No," answered Plu, "this happens at times—the mystery of life taking a different form to continue—not just for yourself but for others. As we say, 'Different strokes for different folks.' You'll go on, Vol. We just need a few more periods of cold and warmth to get rid of some things and grow others. I think what you need is some afterripening."

"Afterripening? What's that?" said Vol.

"Afterripening is inner softening to allow."

"To allow what?"

"Whatever happens. You don't have to like what happens, but if you don't fight it, life is easier. Afterripening creates receptivity."

Plu paused a long time to get Vol's full attention before continuing. "Sometimes we have hardening of the attitudes, which we need to slough off. A good winter's freeze could take care of it for you."

Plu made another lengthy pause. "You want to live from love. See what the movement of your heart is. Are you always trying to fix yourself, or do you genuinely care for yourself? Warm up to who you are. The winter and another summer should do the trick."

"So, you're giving me time to learn who I am," Vol said slowly.

"You got it. That's it in a *seed* shell," Plu confirmed. "There's no guarantee, but it worked for me and many others."

"I trusted you before. I'll do it again."

"Trust yourself. That works better."

They gave each other a big hug as Plu got ready to leave. He felt fresh again, just like when they first met. "I feel good in her presence," Vol said to himself after Plu left. "I guess it's because she's just so good. I hope she comes back."

Plu's words stayed with Vol a long time. "She said to trust myself," Vol recollected. "I know the feeling of inadequacy that overwhelmed me earlier came from doubting. What agony that was. It was doubting that imprisoned me so that I couldn't see any way out. I'm going to trust and see if that frees me."

Time rolled by, as did the seasons. But Vol was more patient now. He remembered his decision to trust. He reached the point of trusting not just himself but life in general. When he thought about himself, he was more gentle. Instead of just looking at himself or trying to work things out, he'd give himself an embrace from the heart.

Periods of warmth and cold came and went. Vol was more acutely aware of them, and he willingly experienced the feelings they caused in him. He knew he needed to cooperate. He thought about all that had happened to him since he set out on his quest. He had connected with the elements, which expanded his view. But Earth told him something that he hadn't resolved yet. "I have to see them not just connected to me but as parts of myself. Well, Plu said to trust myself, so whatever I need is part of myself. Those elements, then, are really part of me."

Another time Vol reflected, "Plu said to stay open. I must be open to the greater until I become it. As long as I can remember, God has been the greatest for me. Does that mean if I don't close, I'm becoming God in some way?

"Maybe God only appears to come from outside! After all, He is the deepest part of me, and there's so much of me I don't know. I think when God *comes,* it is one half of me meeting the other half, the known me meeting the unknown me. In other words, I am God, seed-style."

When Vol found thoughts pleasing, he would hold them and meditate. The charm would often seize him and transport him to another realm. "My quest has less to do with getting things from outside and more with letting them unfold from within. Accepting them and entering them, they become part of me or just me. I have to acknowledge the oneness to let the union be fruitful in me. Plu had said I got it in a *seed* shell . . . "

The words '*seed* shell' exploded in Vol's memory and now, his consciousness. "I got it! I got it! I'm a seed! I'm a seed!" Vol was enraptured. He didn't know how it happened, but he knew a truth had come to him, appearing out of the dark and making everything luminous. He knew that all was OK and nothing could hurt him. There was order and love in the universe and he was part of it. No, not a part of it;

he *was* it. He felt there was nothing that was separate from himself. Everything felt right.

Afterward, Vol tried to give an account of the happening to himself. The experience was literally shattering because at the point of self-realization, Vol's outer shell cracked open and he started to grow. He discovered he was an embryo, or miniature tree, with a protective coat that was a defense, a defense to be broken for his true self to emerge.

He realized that it no longer mattered whether things were within him or outside, his or someone else's, as long as they were present for him.

What he had put to rest was his own fear, the fear of considering something to be other than himself. Nothing was alien. All was his. All was Vol.

Vol thought what primed him for his self-realization was surrender—allowing all to be a part of what he was. He couldn't explain it. He didn't know how all was one, but he knew it to be true as sure as he knew he was a seed. "Maybe I'll learn that in the next phase," he said to himself.

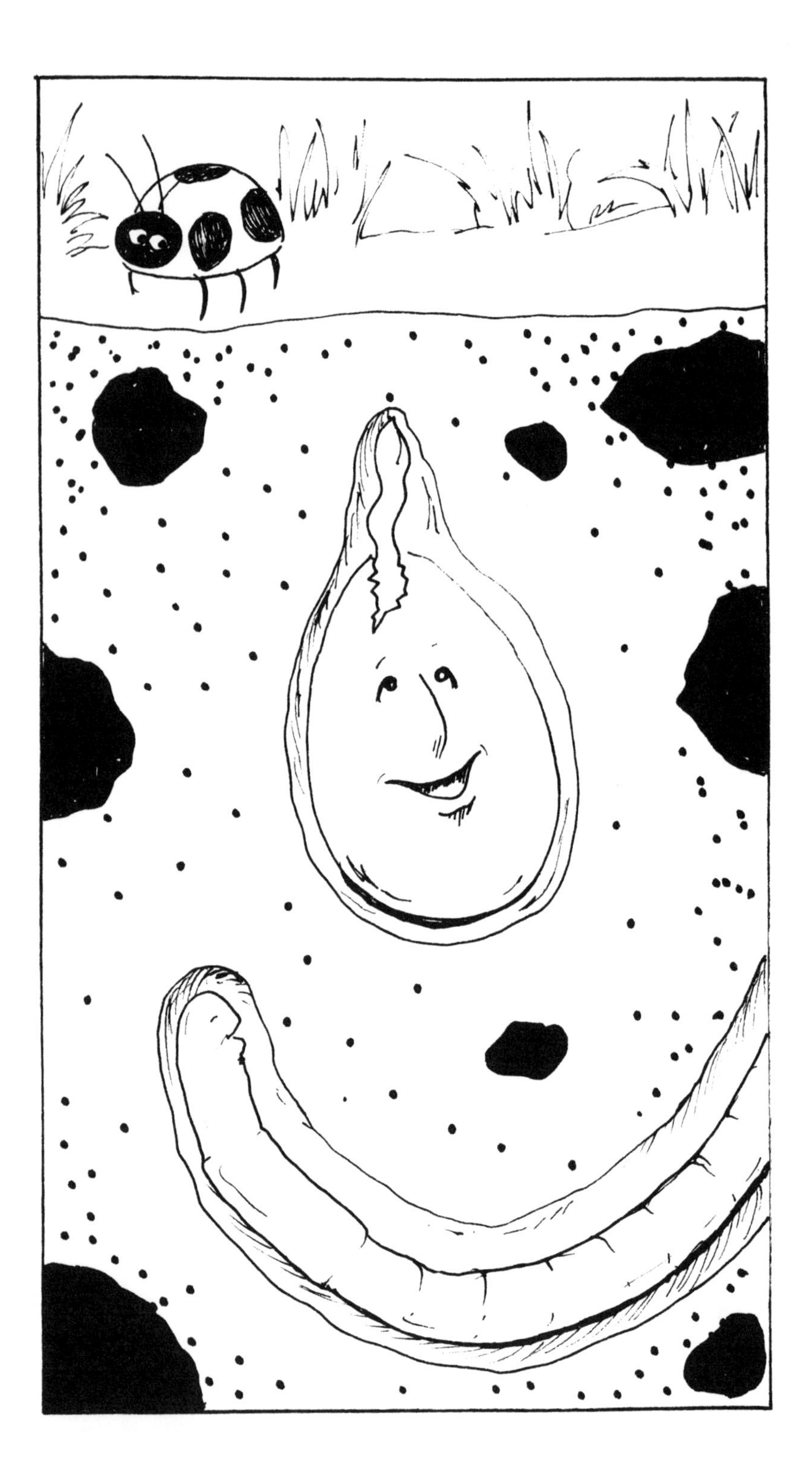

expanding . . .

Chapter Four:

breaking into new life

Once the shell broke, Vol grew fast and vigorously underground. He felt at home. He headed straight for the light that had kindled a spark in him, and he experienced a vibration with the light glowing within him like a firefly. It was an inner power that he had not known before.

Vol expanded in all directions. At one point in his meandering, he ran across his former home, the cracked shell that was now broken in several pieces. He had time to see how the pieces were breaking down and changing in appearance. He wondered if they could feel his new life surging around them.

The light got brighter and the temperature warmer. He had no sense of time passing. Every moment was complete and satisfying. Then a new sensation hit him. He felt a hot flash with a chill inside. He had surfaced. He was above ground and underground at the same time.

The change brought a comforting warmth that spread through Vol. With each moment, he could sense more of himself exposed to light. He had come through darkness to the light. Vol realized that this was the result of his connecting with the elements underground. He never dreamed that his hunger and thirst would be satisfied like this.

Vol grew quickly, bringing forth a leaf, then another, and another. Wanting more leaves, Vol threw out more branches. As underground, he spread in all directions. Vol had become a sapling.

Vol stood on the edge of a forest. On one side were the trees. On the other were high bushes and a few homes in the distance. He felt small with the huge trees around him. He wondered if he could catch up with them. "Everyone has to start someplace," he said to himself, "and this looks as good as any. Come to think of it, it's not my first start."

Vol saw himself taking on the shape of other trees but still felt small in stature. "I wonder if I'll always feel little. Maybe it's because I'm different."

Along with the enthusiasm that Vol welcomed, he experienced loneliness. He could feel ants and spiders and centipedes on his bark. In his branches he welcomed flies, ladybugs and butterflies. At times he enjoyed the heavier tug of a bird. Even as he connected with them, there was a deeper longing taking shape in Vol's soul.

He wanted something that lasted through all the changes. He could accept change because that was part of the miracle of growth, but he wondered if there wasn't something bigger that holds it all together. He could feel the yearning inside and listened to it. He hoped that he would discover what he was longing for.

One day he felt a sensation different from anything he had sensed so far. Something touched him, but the touch was a brushing embrace and he could feel his heart respond.

"What was that?" he asked himself. He felt the touch again and looked to see a speckled fawn nuzzle him with its nose. "How beautiful!" Vol exclaimed. He experienced a presence that for a fleeting moment quenched the thirst in his soul.

This was a new feeling that Vol wanted to explore. He found something that answered a deep yearning. He knew he was no longer alone.

"Does that mean loneliness goes away when someone comes, or does it have to be a special someone? Will I always have to depend on someone?" Vol knew that his head would not give him an answer. He was already aware that he had to experience something to make it real.

The next time the fawn came by, he brushed against Vol the same affectionate way. Vol asked, "What's your name, little deer?"

"Jeremiah. What's yours?"

"I'm Vol. And I'm growing every day."

"You sure are," said Jeremiah. "I've been watching you since you were just a sprig. I felt drawn to you but decided to wait till you got nose high."

"Wait for what?" asked Vol.

"To touch you," answered Jeremiah. "You were so little, I thought that bending down to touch you would put too much pressure on you and maybe hurt you. Now that we're equal, I can touch you."

"You mean you have been watching over me all this time and I didn't know it!" Vol wondered aloud.

"I guess you didn't know, but I did."

"I feel embarrassed, but I do appreciate your caring. Thank you," said Vol. There was a pause. "You say we're equal, and I do feel a closeness with you, but it seems like you have been a big brother to me."

"Hey, we're in this together," answered Jeremiah. "If I have been a brother to you, someone else will be a brother or sister for me next time. There's a delicate thread running through us that connects everything."

"Where did you learn that?" asked Vol.

"My Dad. He's always teaching someone. He's the herd chief and the other deer come to him for advice."

"What's a Dad?" asked Vol.

"You don't know?" replied Jeremiah. "He's my Father, or one of the two parents I come from. But there have been deer who have joined the herd because they were alone and didn't know who their parents were. My Dad says we all have parents, although humans have the most strays."

"Why do you suppose so many humans are left alone?" asked Vol.

"Maybe it's because they smell. That's why we always try to stay upwind of them."

"Hmmm! Well, who's your other parent?" asked Vol.

"She's called Mom or Mother." Jeremiah was surprised Vol wasn't familiar with these names. He felt a pang of sorrow for his friend. "Names aren't that important. Dad says a name is a social tag that doesn't tell who we really are. Our real name comes from our role in the universe. Anyway, my Mother fed me before I could forage for myself. She and my Dad will take care of me 'til I'm old enough to do that. Dad said I came from both of them because they love each other so much."

Vol became thoughtful. He remembered that Earth had said, *"I am the Earth that gave you birth."* He wondered whether his loneliness was a longing for her. In his closeness to Jeremiah, he felt the yearning more than ever. "What does your Dad say about feelings?" asked Vol.

"He says to honor all of them," answered Jeremiah. "It's a way of honoring ourselves. I went to him when I was first attracted to you. He told me to stay with the feeling and warned me what could happen if I just acted out."

"What would happen?" asked Vol.

"Dad showed me a sapling broken by a fawn who recklessly followed his impulse. He said acting out is often a way to avoid the feeling, but if I just live with the feeling it will eventually lead me to the right action or experience I need."

"Did it?" asked Vol.

"Oh, yes," answered Jeremiah. "Dad said respect for my feelings leads to respect for others. My attraction to you was out of love," he explained. "I could never hurt you, so I had to wait 'til you were bigger and stronger.

"Now the new feeling is that we are equal and I have a companion. Besides, now you have more foliage that scratches my face and tickles my nose. It was worth the wait."

Vol was beginning to realize that Jeremiah and he were tools for each other. Jeremiah's words reminded him that his own feelings could be trusted and provided a kind of pathway.

Big brother's touch was magic and now Vol had nothing to fear. Feelings are safe, even holy, from what Jeremiah says, thought Vol.

"My longing is valid," Vol said aloud, "and every time I love what I'm feeling and experiencing, I fill up a little of the emptiness. Loving what I'm feeling takes me into a new experience each time."

"You learn fast, brother. I have to go now," Jeremiah said. "I'll see you soon."

Vol could hardly wait. He saw that both of them—and the herd—and the elements—and perhaps everything—all long for safety and oneness. But he also learned that he had the power to create security for himself and maybe for someone else.

When alone, Vol thought about his conversation with Jeremiah. "I need to act on my feelings the way Jeremiah does. Instead of suppressing them or wanting them to go away, I'm going to use them."

He cleared his mind to enter his feelings with full force. He found new value. "The feelings *are* me. I trust myself when I love them, and I don't need to justify them." By loving the feelings, he sensed a shaping power within himself. He was accepting as his own truth what felt expansive and enlightening.

The practice gathered his scattered forces and centered the new Vol that was emerging.

The next time Jeremiah came, he enjoyed a quick nose job that ruffled Vol's leaves. Vol wrapped a few stems around Jeremiah's ear. They continued their discussion.

"I shared with my Dad what we talked about and he told me what he thought. He believes that longing brings its own fulfillment by letting it be a wave that the heart accepts and enters. The longing stretches him out and reminds him of who he is. By loving his own longing, he brings in the embrace of the universe. Longing connects. In that gesture, he learns his link to what is beyond the world. It's allowing what's present to be his presence."

On hearing this, Vol sensed a direction for his own heart. He thought of his feeling at the moment as a single sound blending with others. It seemed to enter their presence.

It was good to have the deer chief's affirmation. "I feel myself swaying with what you are saying, and I'm reaching to understand," said Vol, "but it seems to be asking an awful lot from feelings."

"Not really," answered Jeremiah, "when you consider that every emotion is a cry to love or be loved."

"That sounds right. Can you explain it?"

"We can experience only so much, then the mind wants to explain," answered Jeremiah. "My Dad says one feeling leads to another and they all lead to love. Every negative feeling, like anger, hate or jealousy, hides a desire to be loved, and every positive feeling, like joy, praise or appreciation, is rooted in love."

When Vol heard "root," he remembered that in his dormant stage he yearned to put down roots.

Jeremiah spoke, "Deer crave apples, and Dad says, 'Love is the apple all want to sample.' I just don't like the explaining part. My experience loses something when I try to put a label on it. I don't think you can experience the same thing twice."

Jeremiah's words brought things together for Vol. The new *magic mirror* that Jeremiah had given him to look at himself had love for its frame and opened out to mystery.

The mystery was tremendous and fascinating. It drew Vol in and made him realize he did not have to understand everything. In trying to understand, he often put his own meaning on things, and that made him afraid.

He now saw that his fear came from his search for those meanings. "If I let go of my interpretations," he thought, "I could overcome fear." Right now, experiencing the unknown brought him bliss. He decided to let the feeling itself guide him and not try to explain it.

Chapter Five:

branching out

The friendship of Vol and Jeremiah grew as they got older. Vol realized the discovery long ago that he was a seed was only the beginning of his journey. The more he talked with Jeremiah, the deeper he realized that life was a continuous creation.

The two friends stimulated each other. Jeremiah told him he loved his family and the herd and wanted to do something for them. But he confided that his goal was to find out what was real in life, and he hoped he could use his talents to help that along.

Vol thought the finest tool was the one that Jeremiah had given him—loving himself and trusting his experiences to guide him. It gave Vol confidence that he could make choices for the kind of life he wanted.

With Jeremiah's love and his new regard for himself, Vol learned to trust himself. He was choosing to be open to everything and hold onto nothing. The resolution challenged his old fears and made him alert to what was happening around him.

He experienced the continuity of life and change and the importance of being in the present. This is what he loved about nature—of which he knew he was a valued part—everything simply is and just happens. All is taken care of without judgment. He saw nature's beauty and order that enabled him to put his life into a larger picture. Vol was absorbing the lessons he thought nature offered. He belonged, and he was learning to just be.

He looked upon a rose and reflected how different and how similar they were! "This beautiful flower appears passive, but it's sensitive to all its surroundings. Its roots not only touch but are firmly embedded in the earth, taking the water and minerals it offers." Vol wondered if the rose had to learn to connect the way he did and if Plu helped her along. "She turns her beautiful face to the sun, and will do so all day long, thriving in its warmth and life-giving rays.

"She welcomes the rain and looks so refreshing afterwards. The whole plant is a living and breathing splendor, exchanging gifts with her environment in a beautiful and unwavering manner." Vol didn't know the whole design, but the harmonious practicality of it all led to his belief that there must be some sort of purposeful plan. "I, too, am alive and enjoying it."

Vol continued his musing. "The rose is in tune with the forces that make it what it is, so different from me, taking what it needs, yet giving its red and perfumed loveliness to the world. It transforms totally foreign elements into a unique creation, one so different from myself. I still wonder what my goal is, yet I know I will find it."

His growing self-confidence attracted others. Squirrels scampered over him. In the fall a particular squirrel stored nuts and seeds in one of his hollows. Vol wondered whether a squirrel like this had carried and hoarded him as a seed. He thought, comfortably now, that it was just another answer that it was all right not to know.

He decided to call the squirrel Scampy. He liked Scampy's boldness. He would go right up to boys and girls and take seeds or nuts out of their hands and run back to Vol with them. The whole thing delighted Vol. When the children came with squirrel food, Vol got a massage from Scampy running all over him. He liked the closeness of the children and the adults who came to watch over them.

Birds visited Vol often and stayed longer, which made him wonder more at the beauty and mystery of life. And beauty brought bliss. The sun amazed him with its brilliance, contrasting with the soft glow of the moon. He saw the changes they brought to himself and everything around him, day after day, night after night. He marveled at the dew spreading on his leaves during the course of the night and turning to twinkling rainbows when the sun rose. The colored bands disappeared just as the stars did when the sun grew bright.

Daylight always brought much activity. One visitor he relished was a red-headed woodpecker. He didn't perch on Vol's branches like the other birds. Although young, this little guy was so tough he could cling to the trunk down low or up high. For a long time he was cautious. With any loud noise, he would quickly sidle to the far side of the tree, reappearing watchful when the noise stopped.

As he got older and more sure, the woodpecker pecked away at Vol's bark to dig out boring insects. He would make holes with his beak the size of a finger, then zap his prey with his long sticky tongue. The lovely bird's closeness reminded Vol of the first time he felt loved by Jeremiah, who was a full-grown buck now. The ants and beetles that got under his skin turned on the woodpecker's searching instinct. Vol loved the *kak kak kak kak* of the woodpecker, and he felt restored by the action.

What moved Vol the most about the woodpecker was his stunning beauty. His chest was snow white with a solid black back. Lower down, the black wings had large white patches. His crown was brilliant red. Vol remembered the bird had a brown head when he first came to him as a fledgling. Vol rejoiced in this miracle of color to the point that he felt it was his own.

Vol's heart was expanding to embrace all. There was so much for which to be thankful. The love he felt convinced him of the presence of goodness. His own wonder transported him. He gloried in the beauty and mystery of being one's self, yet connected to all. He couldn't put it all together, he told himself, but he found a way to satisfy himself.

He turned more and more to poetry in expressing his deepest longings. The paradox was that in venting himself in words and images, he passed over time and space and entered another consciousness. He knew life was not only what he liked to put in words—beauty and bliss, wonder and love, goodness and God—but composing enabled him to bring those qualities into his own life. Into himself, really, thus giving him a new identity. His favorite poem was one he called

THE CONNECTION

A kiss leaves two behind,
  creating one without fear.
Oneness bids me kneel at the altar of myself,
  because there is no I.
I am what I have always been,
  the one self is God as me.
Oneness lies within,
  urging love at first sight.
What is needs no cultivation,
  only blessing.
Earth makes each a light to his world,
  so that all decrease and one grows,
  killing fear with singular vibration.
Intimacy bridges two shores
  to live the oneness between,
God is the kiss,
  his Being my living.

Vol wished he could share his poems. He wondered whether Scampy or Jeremiah or the woodpecker liked poetry.

Chapter Six:

the magic of love

One day, Scampy came running to Vol, leaped into his arms with a load of seed, and quickly climbed to his favorite hollow between a large bough and the trunk. Vol thought this would be a good time to talk about poetry because the little fellow would be clawing and gnawing for a while.

But a strange thing happened. Right after Scampy went up, a boy came along and followed him. This was the first time a human had come to Vol. Vol was surprised at his lightness and how comfortable he felt.

The boy climbed slowly. He would stop often and run his hand over a bough and look out from the tree. About halfway up, after doing the same, he sat down. In the mean time, Scampy had jumped to another tree.

It was then Vol realized that the boy wasn't chasing Scampy. He sat there, leaned back with his head against another bough, and looked out at Vol didn't know what. After what seemed a long time, the boy climbed down and went away.

"Well that was interesting," Vol said to himself. "It looks like the boy followed Scampy but wasn't with him. I wonder what he wanted."

The next day the boy came again, but this time Scampy wasn't around. He climbed to the same spot and sat down. He wiggled his seat and shoulders, leaned back against the bough and settled in.

After a while he took an apple from his pocket and ate it. While he was chewing, he pulled a branch of Vol's white blossoms close to his face and smelled them. He seemed delighted, inhaling the blossoms deeply.

He repeated the action, not hurting the branch or plucking the blossoms. Vol felt a warm glow when the boy did this. It reminded him of his first love, Jeremiah.

The boy often came back, sometimes with Scampy and sometimes not. Once when Vol and the boy were alone, the boy wrapped his arms around Vol after smelling the blossoms, hugged him tight and said, "Thank you, Mr. Tree."

"My name is Vol," said Vol.

"What?" exclaimed the boy.

"My name is Vol."

"I heard what you said," the boy cried, "but trees don't talk!"

"Why not?" replied Vol. "Have you ever talked to a tree?"

"Well, no."

"Have you ever hugged a tree before?"

"No, I haven't."

"Then why did you hug me?"

"I just felt so good inside for what you are."

"Maybe that's all it takes to hear us talk."

"What do you mean?" inquired the boy.

"If you love, I like to call it connecting, you can communicate with anything. You connected with me when you caressed my blossoms and did not hurt me."

"My Dad told me to be careful with your blossoms because they will be apples, and I love apples. That's why I love you and hugged you."

"So, I'm an apple tree!" Vol exclaimed. He couldn't contain his joy. "And I'm to give apples to boys like you?"

"To everybody who wants them," said the boy, not quite understanding. He was surprised enough that this tree was talking. "My sister won't want one. She likes watermelons. She gets my mother to buy one, then we all have to eat it."

"What's your name?"

"Eddy," answered the boy.

"Where did you get the name Vol?" asked Eddy.

"I don't know. I've always had it." The question pained Vol, thinking he had no parents like Eddy or Jeremiah. The yearning in his heart rose again. "Will you keep coming back, Eddy?"

"Oh, yes, I'd love to. Now that I know we can communicate, there's so much I'd like to talk about. I have to get home for supper now. I'll be back day after tomorrow, because we're going away tomorrow. Your blossoms are so fragrant; there aren't any others like them." With another hug, Eddy was off.

"So I'm an apple tree!" Vol repeated when he was alone. "Life is full of surprises. I suppose the discoveries never stop. When they are as wonderful as this, I hope not."

* * * * *

When Eddy came back, Vol already had in mind a subject for conversation. "Eddy, do you like poetry?"

"I sure do. Mother reads it to me all the time, and we study it in school. Mom asked if I would like to try my hand at it. When I said yes, she taught me a few things. Do you like poetry, Vol?"

"Yes, I do. It seems so connected with love. What did your mother tell you about poetry?"

"She said everyone is a poet at heart, and you can write a poem on any subject. She said we all have to express ourselves, and poetry is a good way. Do you write poetry?" Eddy asked.

"Yes I do, and I wrote a short poem recently. Would you like to hear it?"

"Yes," Eddy responded quickly, delighted he could share with Vol something dear to him.

Vol recited his poem, "The Connection." When he finished, Eddy said, "I like the part about a kiss leaving two behind because they are now one. The other becomes me."

"Thank you. You got what I was trying to say."

"I felt we were two, you and I, until we started communicating. Now I don't feel we're separate.

You're telling me that I can communicate with anything, aren't you?"

"My experience tells me I have to become the other. You'll have to find that out for yourself, Eddy."

"I've had the experience of becoming the other," the boy answered. "I just go into my feeling and fall through it. It's a magic game I play. I say the words and then let go."

"What do you say?" Vol asked.

Eddy held his palms against his chest, then turned them upwards while extending his arms. As he did so, he said:

I feel drawn and want to share.
Make us jell so Eddy's there.
I feel drawn and want to share.
Make us jell so Eddy's there.

"Like with the hyacinth at your feet."

"What are you talking about, Eddy?"

"There's a beautiful hyacinth where your trunk meets the ground. It's the only one around. I looked for others and couldn't find a one. I was alone and decided right then we were meant for each other. I sat with her and did my magic and wrote a poem about us. Would you like to hear it?"

"Of course," said Vol.

Eddy looked dwon in the direction of the hyacinth. He paused, breathed deeply, then began to speak:

THE MAGIC OF LOVE

I felt a oneness with the earth,
With the water and the cloud,
But most of all with all the trees
Who now were not a faceless crowd.

Before I plunged into their shade,
A hyacinth in sunshine bright
Laughed at me in morning wind
Its purple petals my delight.

One alone it sang to me,
An ancient hymn of scented beauty,
And I, enthralled, knelt down to her,
As homage seemed a sacred duty.

We swayed together fresh as air,
Silent in a loving dance,
I could only hope that she
Wouldn't break the holy trance.

I sought how I could hold the charm,
For never could I pluck this flower.
What is there that dare would harm
This petaled shrine of healing power?

She nodded in a little bow
To where her shadow fell with grace,
And there I saw a leaf outspread
With holes that marred a graying face.

The leaf did hide in beauty's shade,
To savor well the hallowed ground,
He gave himself to what seemed best
Before he starts another round.

I saw the answer to my prayer
And took the leaf into my hand,
Kissed it fondly for his gift
Of bringing light to shadow land.

From a coupled harmony,
Grew a larger symphony,
Deep within an empathy
Saw my comrade change to me.

In a sunset or in ocean
I had sensed a living song,
But most of all inside I heard
Beauty chanting all along.

Smiles of God that foster dreams,
His genius forms a holy ploy
To keep us young in mind and heart
Beauty shapes each girl and boy.

When Eddy finished, he said he loved the purple hyacinth so much he wanted to take it with him and keep smelling its perfume. "But I had become one with it, so I didn't need that. Instead, I took one of its dying leaves as a personal remembrance."

Vol didn't answer. He was crying, overwhelmed with the love he felt for Eddy.

uniting . . .

Chapter Seven:

food

Time passed. Vol and Eddy spent much of it together. They talked, laughed and shared their poems and ideas.

Nature nourished them. She lifted their spirits by imposing a greater rhythm. Her continuity amid change reminded them of their respective journeys, of letting go and opening to new phases. Birth and death. Sun and darkness. Sweat and ice. The elements furious and flaccid. Beauty and decay. Woodland fragrances and putrid odors. Pine needles and cottony fluffs. They felt an equilibrium coming from nature that gave them power to expand. The abiding presence of her beauty enhanced their lives.

To them, the world was a friendly but mysterious place, and the big mystery was self. That made life an opening to the unknown. Knowing this, they agreed, clarified life and gave it focus. They both liked to express this in their poetry.

In one conversation, Eddy said, "I think I can be anything I want to be, but I just don't know what I want yet."

Vol asked, "Isn't that why you go to school, to prepare for the job you'll have?"

"Yes, that comes later and I'll have to decide. But I want more than just a job. I want to be something, like a great leader or a genius who creates. I feel like the world is mine, and I want to share with others what I have."

"That seems like a noble idea," replied Vol, "but perhaps a leader doesn't necessarily have followers. To me, a leader is a genius—someone who can create himself and fulfill his destiny."

"What is our destiny?" asked Eddy.

"It's for each one to find out."

"How does one do that?"

"He lives, loves and learns."

"You make it sound simple."

"Simple to say," Vol said, "but it may be hard to do. Learning is creating."

"Go on," Eddy urged.

"Real learning changes us. It creates us. I see learning and creating as two sides of the same coin."

After a quiet moment, often normal between them, Vol continued, "The leader who is learning becomes more of himself, so to speak, by joining what happens to him. It may be painful, but he doesn't resist. It's another form of connecting, which we talked about before. It's his ingenious way of fulfilling what is deepest in himself. I think everyone can do this. Our feelings and longings direct us."

"But we have so many feelings and longings," said Eddy.

"Only on the surface," answered Vol, remembering the wisdom of Jeremiah. "Our wants are based on our needs, and the greatest need is love—to love and be loved. Until something becomes a heart-swell, we're not whole. Love is what fulfills us."

Vol and Eddy shared their friends as well as their ideas and poetry. Scampy was already their common chum, and Eddy met Jeremiah. They admired the woodpecker together, especially his scarlet head. They decided to call him Red, and hoped someday to involve him in a conversation. If only he weren't so flighty.

Eddy would sit for long periods, watching the squirrels and chipmunks below him, and the birds and ducks above him. Vol had a large canopy now, so Eddy could see only portholes of the sky. When a flock of geese flew over, he caught fleeting glimpses and could hear their honking from his perch. By the end of summer, Eddy had worn a smooth and shiny spot on Vol's bough and trunk where he sat so often.

One day, late that summer, Vol found himself covered with red blots. He thought he might be sick, but he felt good and healthy. Then he thought maybe Red invited his relatives for the weekend, but they weren't bobbing around as Red usually did.

Finally he saw that the blots resembled the red apples Eddy ate. It was then he realized that he was sporting his first crop of apples. It made him happy that he would have something to give Jeremiah and Eddy.

The weather that spring and summer blew in seeds and spores on strong southerly winds. The dampness and heat of the season, in turn, made for abundant growth. Vol had not seen bushes like the ones around him. He knew that when they died or went dormant for the winter, his roots would contact their remains and perhaps use them for nutrients in the spring. But right now, something told him he needed to be careful.

Chapter Eight:

too high

In spite of the warnings Vol gave himself through the winter, when spring arrived, he couldn't wait to sample the new food available. The wild taste excited him, and he wanted more.

He could feel the lift in his limbs and branches all the way out to his leafy fingertips. He wondered whether this was good for him, but the new sensation was too powerful and he devoured the foreign substance.

Sometimes he thought he would explore these new and strange feelings. At least he should ask his friends or someone familiar with them. But the adventure was heady, and he never took the time. That summer one of Vol's limbs grew out of proportion to the rest of him, and it towered above the neighboring trees as well. Vol felt proud to be so tall, and couldn't see anything wrong with it.

Jeremiah and Eddy told him he looked strange, but Vol didn't want to talk about it. The two friends wanted to help, but there was nothing they could do.

Storms were normal in those parts, particularly in the summer and fall. Late one afternoon when the heat had built up, heavy clouds moved in and a strong wind whipped through the trees. Thunder and lightning warned that an electrical storm was coming.

It only sprinkled at first. Frequent bolts of lightning lit up the dark sky as thunder rolled through the forest. Then, when the sky was darkest, a bolt of lightning struck Vol's new, high limb and caught it on fire.

Soon the limb blazed furiously. As the fire burned down the single limb and threatened the entire tree, it started to rain harder. Then the fire reached the hollow that Red had used for eggs. The fire could not jump the gap. That, plus the rain, caused the fire to die down and finally extinguish.

As the last embers of his limb turned wet and gray, Vol felt the presence of an old friend.

"Plu, is that you?" asked Vol.

"Yes," answered the raindrop. "I brought help. You're so ungodly big that I thought you needed a dousing."

"I certainly did. If you hadn't come with all your friends, I may not be here. How come you all went crazy?" asked Vol.

"What do you mean?" asked Plu.

"Light, heat, air and water were my buddies before, and I learned to connect with you underground. But all of you sure acted it up during that storm and almost killed me," responded Vol.

"I guess you see how anyone can be a demon at times. Speaking of demons, how the devil did you get one limb so far out of line? I meant that about you're being ungodly big. Hellish ugly, too, I might add. That's really what almost destroyed you. What happened to you?"

Vol felt sheepish at the question. "It was something I ate," he said, hoping that would end the subject.

"You must have eaten some bad stuff," said Plu. "What you eat has to be good for you. I passed over a number of times and you sure looked ugly with that limb sticking out like that."

Vol could feel Plu searching. "I got mixed up and gave in to a craving," Vol confessed.

"You forgot who you are," replied Plu. "Remembering is recreating. It's good when you're confused to find that silent space within to appreciate yourself."

"Do you mean entering the silence like you told me when I first got started?" asked Vol.

"No, no, no, that was baby silence," answered Plu. "I'm talking mature, adult silence. Look at that chipmunk down there. He's been around a long time. He squeals and scurries, goes into light and shade, eats and explores, knows and loves, but then he goes into the silence of his hole where all is quiet and he just is.

"Just be," continued Plu. "At times you have to listen to the silence to understand who you are. Real knowing is not an end but an awareness that there is always something more. You are not only a trunk but also branches reaching out. You have a particular gift because you are a totally unique being. No one ever existed like you and no one ever shall. You have to be beautiful. Remembering who you are shapes you for what you want to be."

Chapter Nine:

reunion

The electrical storm that almost killed Vol resulted in a forest fire that destroyed trees and bushes close to Vol and throughout the forest as well. It even damaged some of the houses nearby. The devastation meant little food for the animals the following season. But for Vol, the burned-out space around him permitted him to grow in all directions. This time he kept his balance. He became such a large and beautiful tree that even Plu praised him.

It wasn't so fortunate for many of the animals. As the summer progressed and the food supply dwindled, many of them died or fled to other regions. Scampy and Red had escaped the inferno by fleeing beyond the fire line. They were only too happy to return to Vol, one of the few healthy trees around.

Vol gladly welcomed his old friends. When Eddy came and climbed to his usual place, it made Vol's joy almost complete. The companions would share as they used to. What Vol thought about now was

Jeremiah, and he asked Eddy if he had seen him. Eddy said he hadn't, but he would look for him.

Eddy looked far and deep into the burned-out woods and beyond their charred borders. He returned to the woods many times, each time searching a different section. He found no trace of Jeremiah.

When Eddy told Vol and his friends, they said they had been searching too.

Scampy then spoke up. "I've got good news and bad news. The bad news is that I had given up Jeremiah for dead and went looking for dead deer. And I know how to look. The good news is that I didn't find any. So I think he's out there somewhere." All had been afraid to admit the possible truth. Scampy's news was welcome, for they had been fighting back tears.

Eddy had many troubling dreams and sometimes dreamed about Jeremiah. In one, a group of campers built a bonfire that virtually wiped out a colony of ants. One of the few ants who escaped the fire left the survivors to go and search for food. After a long departure, the ant returned dragging a huge fish thousands of times bigger than he was. The ant-hero invited all the ants he could find, even those who had been unfriendly. They had a feast, and the fish provided food for a long time.

In ant-time, it was long enough for the few remaining ants to go out and establish new colonies. They eventually became more numerous than they were before the bonfire.

* * * * *

One day in the early fall Eddy was returning from school. As he got near his home, he saw a thin and straggly deer in the garden, digging among the rows where sugar beets had grown. When he got close, the deer brushed against Eddy in a familiar way. It was then that Eddy recognized Jeremiah. He threw his arms around him and hugged him and cried.

Eddy knew that Jeremiah wanted to see Vol, so they walked to their old rendezvous. Another deer joined them. Further away from houses and into the brush, more deer came to their side. In what was left of the woods, a whole herd of deer fell in behind them. Eddy saw that they were very thin and weak.

He felt sorry for the many starving deer. When they reached Vol, they cried as Jeremiah and Vol embraced each other. The herd came under the tree's branches to eat the apples that had fallen. There were hundreds of apples that littered the earth, many old and soft but just as many fresh and juicy. Every deer could have as many as he wanted. While they were filling themselves, Eddy climbed to his favorite perch, plucked a fresh apple and ate it while watching the deer below him.

During the feast, Jeremiah and Vol talked. As before, they laughed while they enjoyed a serious conversation. Vol asked Jeremiah how he made it through the fire.

Jeremiah said, "I had three wishes in my life. The fire made me see how the three come together."

"What are they?" asked Vol.

"I wanted to discover what is real. For me, reality is connecting so deeply that all is mine and all is myself. Who I am keeps changing. Whatever I join is part of me forever. Eating an apple, the apple becomes me. When we were starving, someone would find a root or a branch that had not been burned clean through. Before eating it, he would bring the weakest deer with him to share it. That formed a oneness that wasn't there before. We weren't just caring for each other, but actually became each other in order to survive."

Vol said, "I've missed you, old friend. And I see you have made wonderful discoveries."

"Secondly," continued Jeremiah, "my great need was to express my unique talents, and I think I did that by leading the deer. I knew you were here, although difficult to find. The fire had destroyed many smell marks, and it was hard to continue since we were weak and hungry. I urged them to keep on, promising I would take them to food. I think my love for you directed me more than anything, and I sensed you had survived the fire. Besides, I just love your delicious apples."

"You already know I love you," said Vol, "but now I admire you and think you are a great leader, not only of others but of yourself."

"The last wish I am now fulfilling."

"How is that?" asked Vol.

"To serve my fellows, whoever and whatever they are," answered Jeremiah, looking around at all the deer who followed him to the giant apple tree. "They will survive because of you, but also because of me. That is very fulfilling. I don't seek answers anymore. I have found I am what the universe is to me—a wonderful opening mystery. How about you, Vol? It looks like you survived the fire in robust fashion."

Vol said, "I never stop learning, and I like that. It helps me create. I had to learn to extend myself by connecting to what is present, first as a seed, and then through my leaves and limbs.

"I find that love is extending myself. The 'I' that's me is not enough. What good is a trunk alone? I need a broader base and so become an 'L.' The 'L' branches out in an Offering of fruit that also brings Visitors, so LOV is born and continues. I have written a poem about it. Would you like to hear it?" asked Vol.

"Of course," answered Jeremiah.

Vol recited his poem:

Loving one is loving all,
Till we see how short we fall,
And grow from there to be so tall.

When Vol had finished the poem, he said, "I once was tall, but ugly. I had to learn to be tall and beautiful. That's what I mean about learning and creating."

When they all had eaten, Jeremiah went to the edge of the herd and talked to the largest buck of all. Jeremiah then led him back to Vol and presented his Father.

The ancient deer's antlers were so wide he could easily hold Vol between them and hug him. The deer chief bowed, scraping his head against him, thanking the great apple tree for saving his herd.

The regal deer straightened up and said, "You know, your name is Vol, which is Lov spelled backwards. That is who you are—LOV."

He then lay down on the ground with his back against Vol. Before resting his head on the apple-rich earth, he turned to Vol and said, "LOV is the apple all want to sample."